I0478503

Silence of the Bow

and other primitive skills

James Willer PhD

ISBN:1542690579
ISBN-13:9781542690577

DEDICATION
My father

CONTENTS

ACKNOWLEDGMENTS

No one goes through life alone
So I would like to acknowledge a few thousand
people, here is just one.

Silence of the Bow

A research project

Many archery hunters of white-tailed deer see a defensive movement on the part of the animal deer prior to impact of the arrow. This movement is caused by sound generated from the release of the arrow from the bow. The sound warning causes a reaction in the survival instinct of an animal. The sound generated by shooting an arrow can be produced by the arrow, string, bow, and the release. This study looks only for sound variation produced by different properties relating to the arrows feathers. The basic idea of this project is to record sound produced by arrows as they leave a wood bow. There are several factors affecting the production of sound relating to arrow's shaft material, bow rest construction, fletching style, and feather type. The project's proposal is to define both objective and subjective sound variations produced by changing arrow qualities. Several arrows shot from a wooden bow and recorded electronically as well as individual subjective input, will establish basic sound variations from the arrow structure. The parameters of this study are limited to arrow sound generation. The study recorded acoustical sound generated as the arrow crossed the rest, which varied due to fletching material, shapes, and style. The recording equipment used was designed to record basic acoustical waveforms, which could then be used to compare how feather stiffness, length and style affect sound production.

Variations of Arrow Construction

It is not defined which arrow produces higher or low pitch as well as decibel levels produced from variations of arrow construction. Using local materials for arrow construction such as turkey feathers and river cane, is there a statistically significant sound change related to the combination of arrow construction technique? This study primarily relates to the fletching.

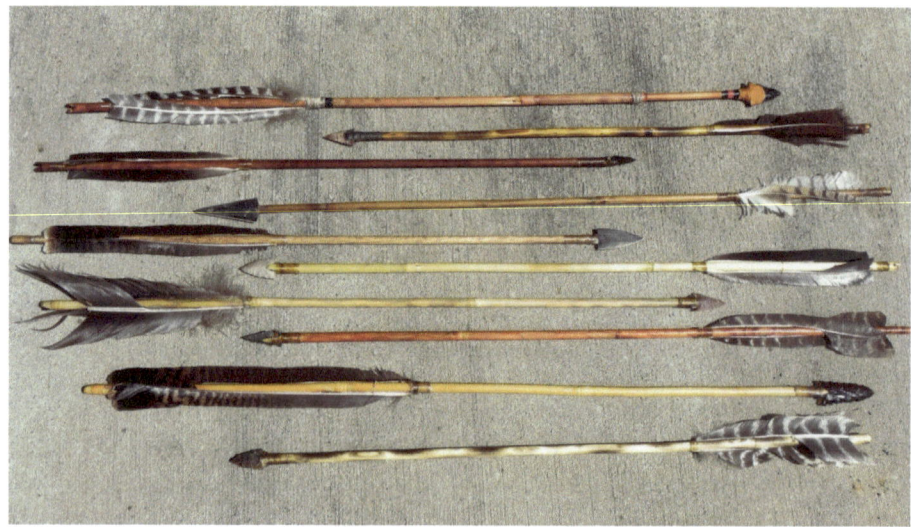

1. Type of fletching design – two feathers or three, helixes
2. Type of feather – primary, secondary
3. Wood vs. Cane
4. Method of fixation of feather to shaft – glued, tied, sinew or natural cordage

Terms Used

1. Decibel – loudness
2. Pitch – Frequency
3. Resonance – Vibration energy
4. Acoustical – Propagation of sound
5. Timbre – Tone (was not addressed in this study)

No prior research was found that clearly defined which arrow is the most effective for hunting any given specific types of game. Without using the most effective tool available to collect food, a hunter and his family could unnecessarily starve. It is in the best interest of a hunter to use equipment designed to be the most effective. If sound produced by an arrow going across a bow rest can affect hunting success, the question arises would primitive man have adapted to use the most effective arrow? Simple methods using local materials constructed in such a fashion as to produce the most effective arrows could be assumed. If low volume sound from the arrow is important, then this would be reflected in the arrow construction.

Methodology Research Design

Using a standard wood recurved bow with a standard pin style arrow rest, several arrows were shot and grouped by construction variations. The pin rest was chosen based on an attempt to minimize secondary noise. The pin rest provided the least amount of contact to the bow surface. This type of pin arrow rest was felt to provide for the maximum of sound generation from the arrow yet minimizing sound produced from the rest. An Apple iPad 2 was used for sound wave recording, and sound wave application software was used for data display. The built in microphone of an iPad 2 was used for sound input, Apple does not provide microphone specifications. The iPad was held 12" from the bow on the side of the arm that was holding the bow. Each arrow wave form was recorded and compared.

In addition to this data sampling, two individuals were seated five feet to the side and five feet forward on each side of the bow shot. Their backs were to the shooter and they recorded their impressions by using numbers corresponding to the control arrows. These two recorders were not able to see the arrows being released. No two hand-made arrows are exact, but arrows were grouped by common design. Only one main variation existed between each grouping. Example of categories: fletching style, shaft material, stiff or soft feathers, method of attachment of feather to shaft. By comparing sound wave forms with the subjective results, a correlation between construction methods and sound should occur.

Flight Groupings

Flight Group #1

Rough hand fletched with turkey feathers, most around five inches in length.

Flight Group #2

Eastern Catawba style, two feathers fletching, some longer than seven inches.

Flight Group #3

Shorter Native American Sioux buffalo arrows with seven inch stiff fletching but low profile, smooth transitions.

Flight Group #4

Control group with modern arrow and one specifically designed with high quill shelf.

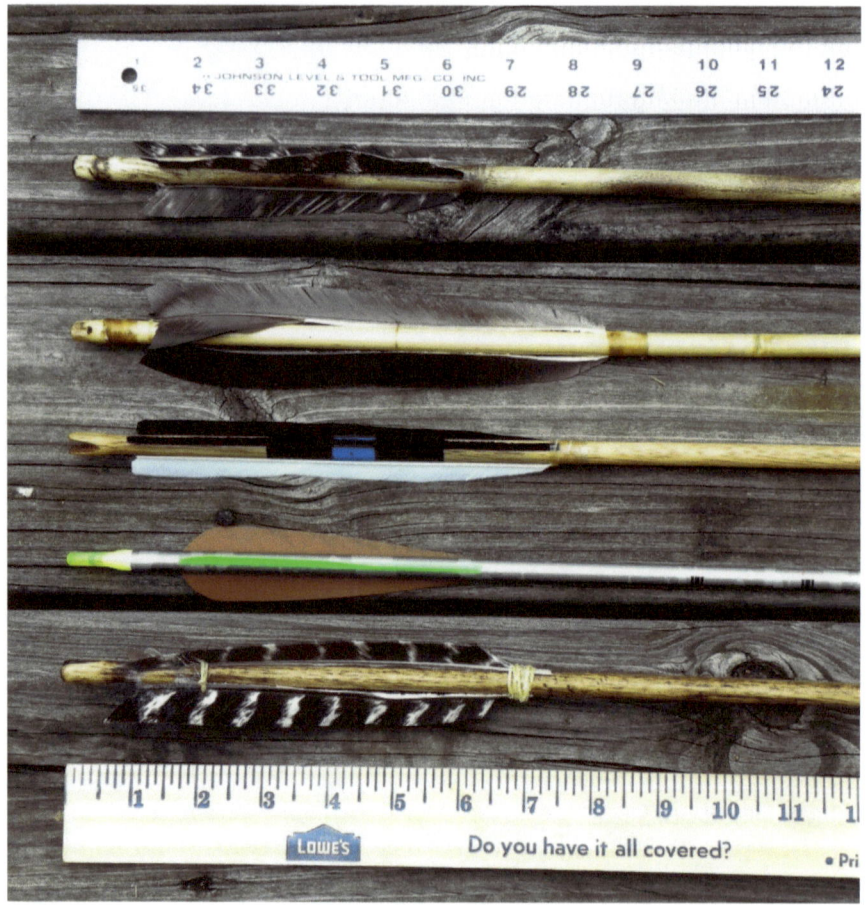

Data Analysis

No personal subjective data was obtained of reliable quality. The sound generated from most of the arrows was not discernible to the human ear. Sample speeds showed an average of .03s in which the arrow sounds were generated. The average human ear could not reliably separate sounds in that short period of time. Only the arrow producing sound discernible to the human ear was the modern arrow. These arrows had a specific tone difference than the primitive arrows. All other differences between arrows were defined though the

use of technology.

This sound wave formation was created by a modern arrow. Note the quick rise and fall of the maximum sound propagation. This is the string noise at arrow release.

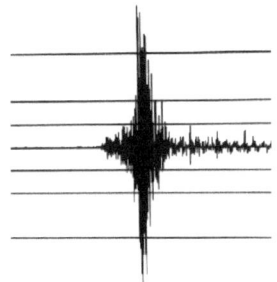

This wave form was created by a primitive style wooden arrow. You can see a smooth transition in sound as the arrow moves.

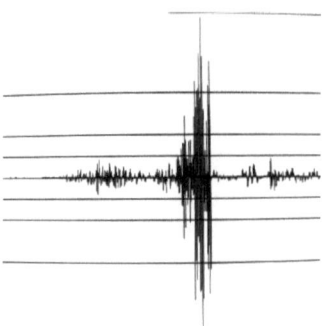

This wave form contains artifact noise created by a rough arrow shaft as it transitions over the bow rest. Remember the artifact is to the left of the peak sounds.

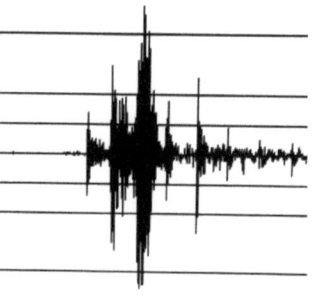

On any primitive arrow there can be defects in the wood, feathers, or wrappings. As the bow rest strikes each one of these micro bumps a sound is created. This wave form reflexes the roughness within the workmanship. T strike of the front wrapping holding th feathers in place is clearly seen.

The wave form expanded out in time shows sound transitions from the shaft noise to the strin reverberations. Peak decibel levels are represented by the parallel lines indicating volume.

Summary, Conclusions, and recommendations

By comparing data, a pattern emerged showing that construction methods affected the sound of an arrow when it is shot from a primitive bow. Different feather patterns create distinctly different waveforms patterns. The dynamics of these changes became insignificant when compared with the sound from the bow string. The results did show that sound produced by impact with the bow rest can be minimized by attention to style detail during construction of the arrow.

A modern arrow rest was used to eliminate as much contribution noise from the rest. This was the arrow rest used.

The average time across the bow rest where sound specific to the arrow was detectable, varied between the ranges of .02 to .05s. Average time for an arrow to reach the target was .5s with the target located 25 yards away. This allowed for a mathematical bow generated average arrow speed of 150fps. Shafts made from river cane reflected sound changes when the nodes or bump nodules on the shaft to strike the bow rest.

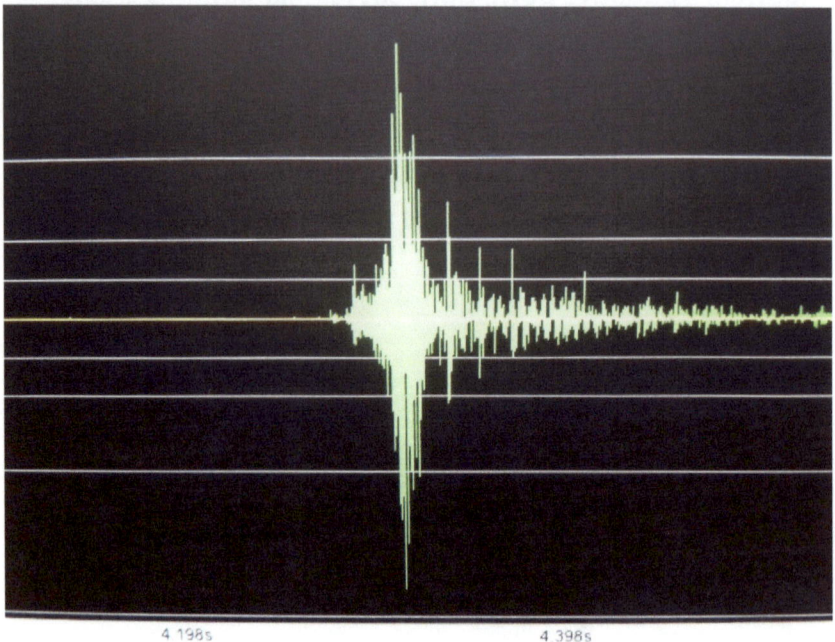

4.198s   4.398s

 The first sounds noted were the arrow shaft scraping over the bow rest pin. It was not found to be the string finger release. Feathers account for 1/3 the total decibel level defection. A heavy fletch shelf or quill tie down that is significantly elevated and does not create a smooth transition off the shaft creates a bounce effect. Rather than this effect being troublesome related to sound, it seems to affect accuracy.

Stiff feathers raised decibel levels, but even stiff feathers of extra length did not increase sound greater than 1/3 the string sound level. As a side note, the bow shooter stated he felt that the arrows from group #3, arrows made of stiff longer feathers, were the most accurate.

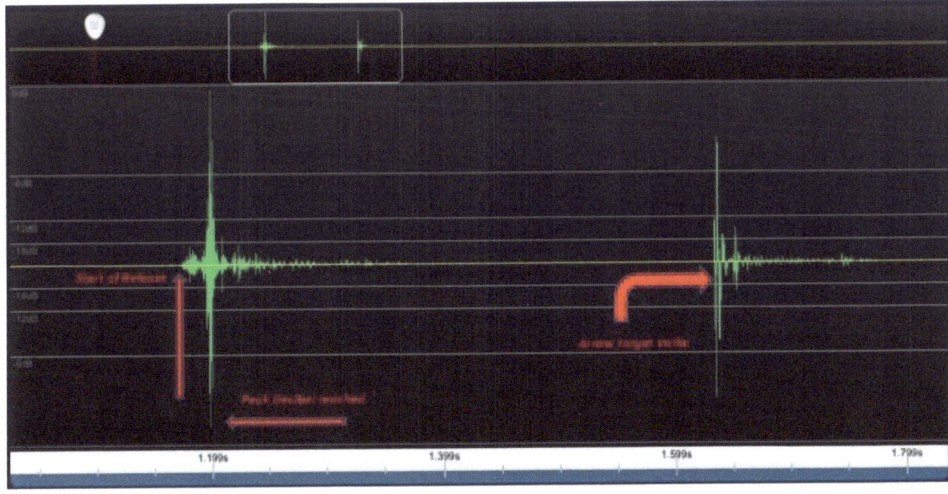

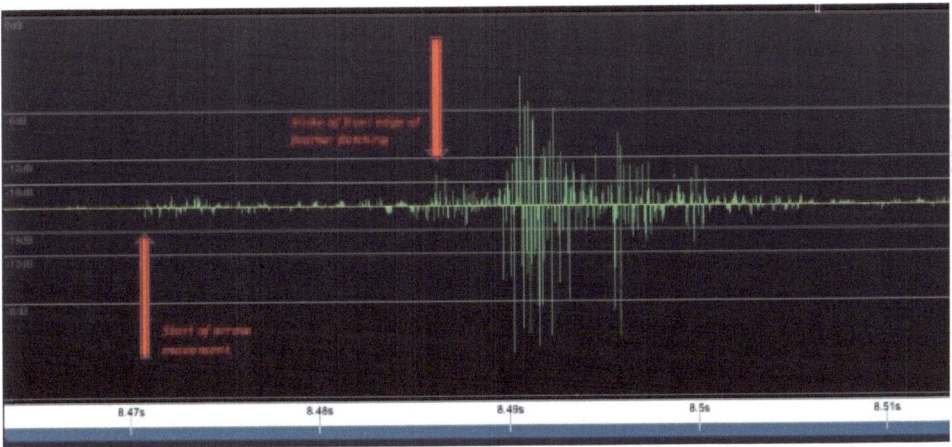

Sound waveforms show pitch differences between wood and aluminum arrows. This is reflected by compression of the waveforms. The feather length can be seen in the sampled waveform by time comparison of feather generated sound.

The total arrow sound accounts for 1/3 of the peak decibel level, when compared to string resonance. A significant muting of string sound was noted when using a heavier arrow. The heavier arrows, even poorly made ones, did not have an elevated decibel sound problem. The string noise relative to arrow weight appears to be the important factor in generation of sound volume. Reverberation is noted after the arrow has left the bow, but this sound is only 1/4th the decibel volume when compared to the arrow leaving the string. End string reverberation noise occurs after the arrow leaves the bow. When taken in total, the main significance is that the arrow weight muted total peak decibel levels. Ballistics and arrow speed would be affected by a heavier arrow but neither of these factors is addressed in this study.

This study looked at the use of wooden arrows and primitive style bows. For effective use of primitive bows and arrows, an average maximum distant to target is within the 25 yard range. Modern equipment has faster speeds and a longer effective range up to twice the distance. Primitive equipment requires the skill and ability to hunt within effective range and close the distance to the target.

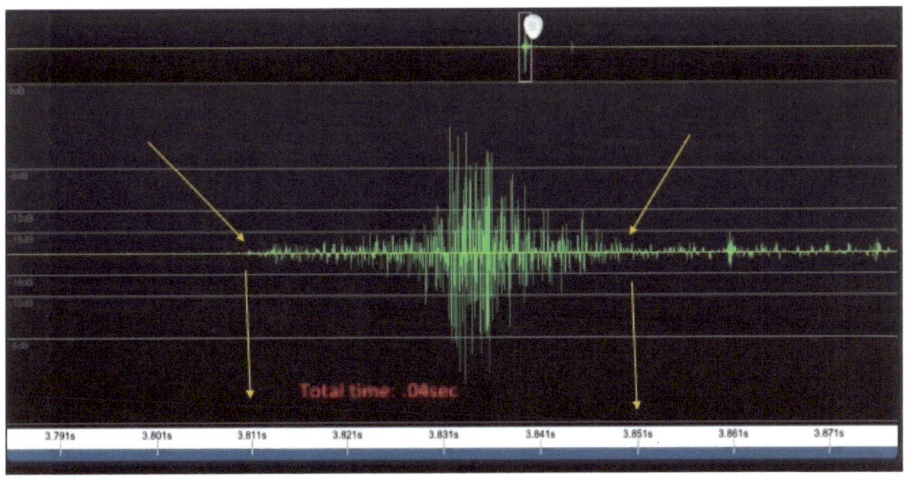

Total time: .04sec

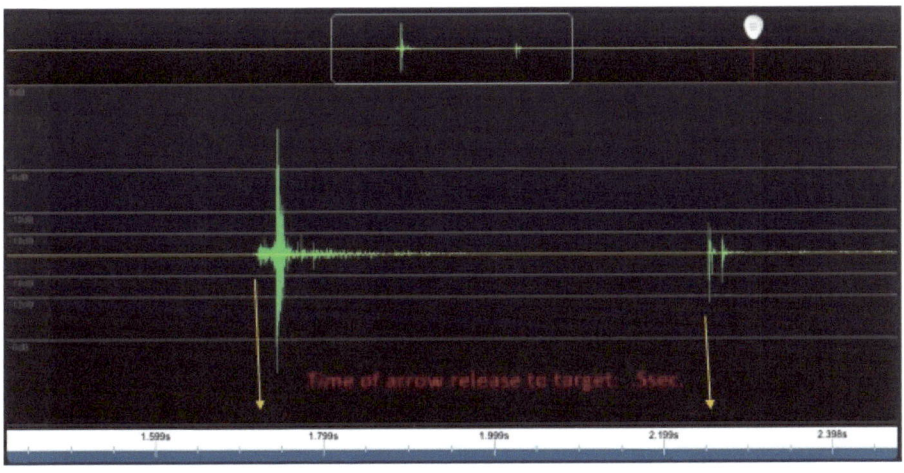

Time of arrow release to target: .5sec

This study was not designed to address modern equipment such as aluminum arrows and compound bows. Modern equipment was used as a sound control measure to assist in establishing a baseline of evaluation. A follow-up study would answer the question: Do string silencers or pads on bow limbs effect peak sound generated by the string and to what percentage?

Research Study: Evaluation of Bow String Sound Related to the Application of Fur Strips into the String

On the great plains of Iowa and South Dakota, my brother John has been hunting deer with a wooden bow for over 30 years. His average distance from bow to deer has been 10 to 15 ft. My other brother Jay, a crippled up farmer has been resigned to blind hunting with a modern crossbow. He has seen a deer jump from the crossbow sound before the arrow struck. His distance from bow to deer has averaged 75 ft. or greater. My last two deer have been taken under 20 ft., using a handmade Native American (Sioux) style bow. Our family style of hunting has been ground stalk and ambush as it was the preferable method taught to us by our father.

My informational take away from this family life experience is that sound travels five times faster than the arrow and as distance increases, the deer is more likely to react ahead of the arrow, no matter what type of sound is generated. Also, if two non-professional, traditional wooden bow hunters get the vast majority of their deer at less than 20 ft., one would assume primitive man did the same.

If the presence of a human being meant death to a deer, the difference between speaking and yelling would not change how fast the deer would jump for safety.

Deer are adaptive to changing hunting conditions in order to survive living in areas with heavy hunting pressure. As modern shooting distances increased; changing the technology, sound, and feel of a bow can became important to the hunter.

Changing sounds from high pitch to low pitch as distance increases, may allow the bow sound to blend with normal ambient noise levels. Using the difference in modern shooting distances, changing the pitch of a Dacron string appears to be a modern problem. Rawhide and sinew bow strings have a lower pitch sound at arrow release than Dacron strings. One method to mute these sounds generated by the string at the release of an arrow is to weave fur into the strings at both ends of the bow. These form small round balls of fur approximately six to eight inches from the bow tips.

The question comes up, do these fur strips on the string affect the pitch and decibel levels of a modern bow string and can these items create significant changes to primitive strings as well?

Relating to a sounds pitch, and human perception, as an analogy, let us look at purchasing a golf club. Given three clubs with the only variable being the sound of the ball strike changing from high pitch to low pitch, golfers will tell you the club of a certain pitch sound feels right. A deaf man cannot discern a feel difference between clubs. This may also hold true with bow shooters and appears to be unrelated to projectile outcome. Changes to the bow string sound may affect perception and bow feel by the shooter, not necessarily arrow success.

Staying within the golf analogy, success is related to physical technique, mental attitude, and technology. As a golfer gets more proficient, mental attitude becomes more important for success. Technical equipment changes only buffer the mistakes made by the shooter, having little impact on the game outcome, and small if any overall performance improvement. In other words, changing to another company's golf clubs probably will not dramatically improve your golf scores.

There is the possibility that the attempted muting of string sound to change the pitch of the sound into a more favorable tone may be perceptual. Noise created from a modern Dacron string maybe within unfavorable pitches which when muted, the archer perceives it as less noise. As an example, it was shown in our prior study that a modern aluminum arrow had a higher pitch sound than wooden arrows but stayed within similar decibel levels.

Research statements

Bow string fur strips change perceived pitch or resonance.

Fur strips change string pitch, but do not reduce decibel levels significantly.

Terms Used

1. Decibel – loudness

2. Pitch – perceptual frequency, to scale high to low, we perceive pitch by the speed of vibration on the basilar

membrane in the cochlea or inner ear. (Hertz)

3.   Resonance – Vibration energy, certain frequency amplitude or oscillations

4.   Acoustical – Propagation of sound, environmental influence

5.   Timbre – Tone (was not addressed in this study)

6.   Damping - Dissipate kinetic energy/loss cycle to cycle, preventing oscillations

Methodology Research Design

A sound studio was available for conducting this study, but leasing costs were greater than any benefit of gathering a cleaner audio sample. Instead, the sampling was done in the same acoustical environment as one would expect to hunt in. Elements we will be working with include pitch, resonance, and damping. Minor changes in these qualities will be discounted as insignificant.

Using a wooden self-bow, two strings were obtained, one standard Dacron and one slightly larger hand twisted string. They were made to fit the bow at a similar brace height. Two wood arrows were shot using each string. An Apple iPad 2 was used for sound wave recording, and sound wave application software was used for data display. The built in microphone of an iPad 2 was used for sound input, Apple does not provide microphone specifications. These two samples provide the baseline for comparative changes. Beaver fur strips were then applied to each string as well as modern rubber silencers made by the company Allen. These devises were placed approximately eight inches from the string ends. Using the same arrows as before, two more sound recordings were made with each string.

Sample Order

Recordings

Feb. 22, 2014

1. Recording 1; White string without silencer\ 7.5 in. brace height

2. Recording 2; White string without silencer\ 7.5 in. brace height

3. Recording 3; Black string with beaver fur as silencer\ 8 in. brace height

4. Recording 4; Black string with beaver fur as silencer\ 8 in. brace height

5. Recording 6; Black string without silencer\ 8 in. brace height

6. Recording 7; Black string without silencer\ 8 in. brace height

7. Recording 9; White string with rubber string silencer\ 7.5 in. brace height

8. Recording 10; White string with rubber string silencer\ 7.5 in. brace height

9. Recording 11; White string with rubber string silencer\ 7.5 in. brace height

10. Recording 12; White string with rubber string silencer\ 7.5 in. brace height

Data Analysis

A comparison was made of each flight group looking for any change and effect from the application of each vibration dampening devise.

This was the bow used for this study. It was hand made by the author.

Sting silencers used

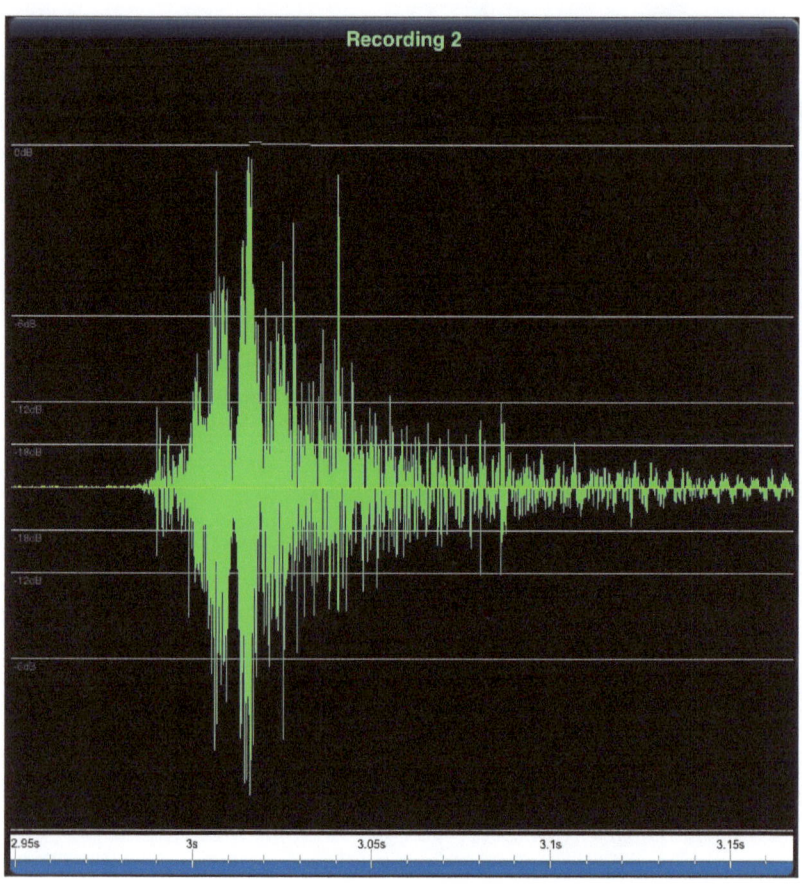

This was the recording without sting silencers

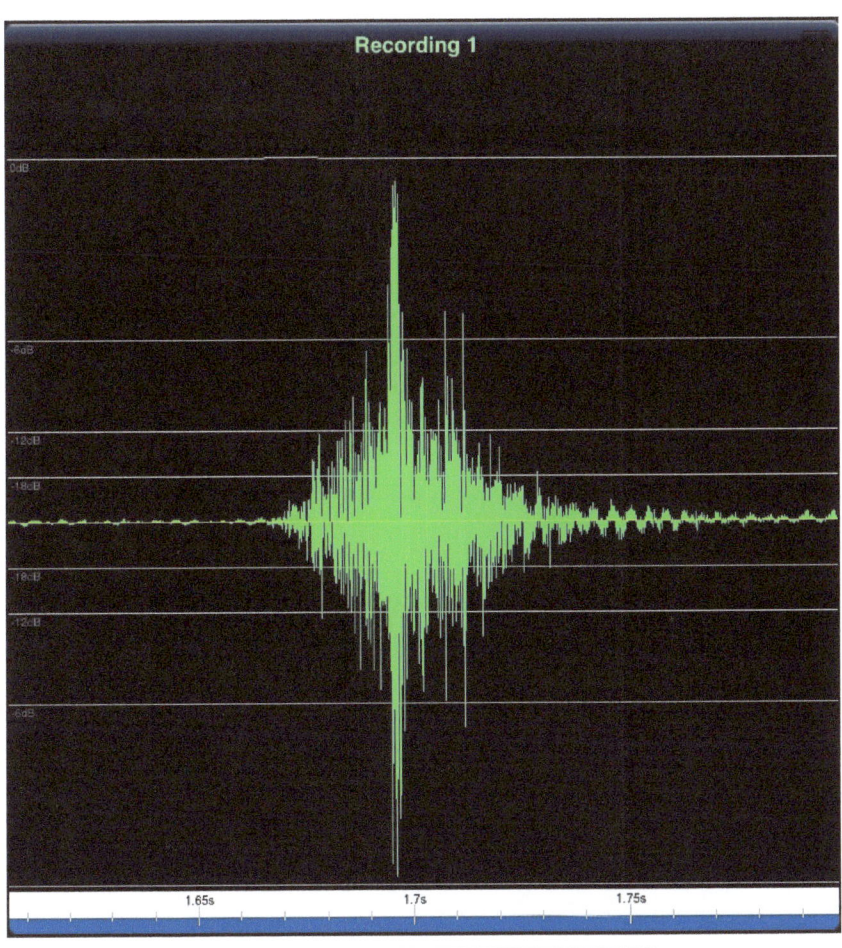

Recording 1

Recording with the use of silencers

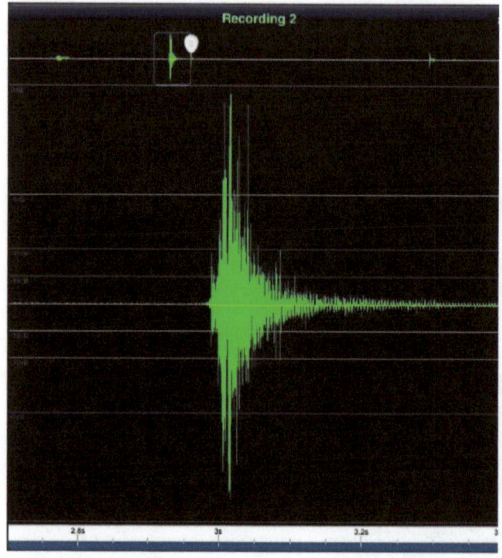

Waveform without the silencer

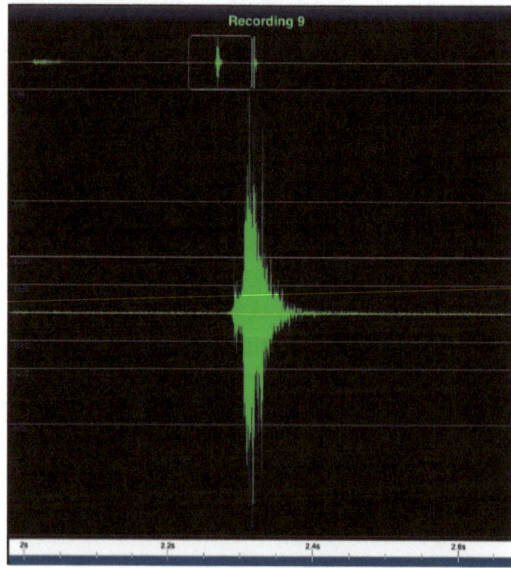

Same bow and string with the silencer

Summary, Conclusions, and recommendations

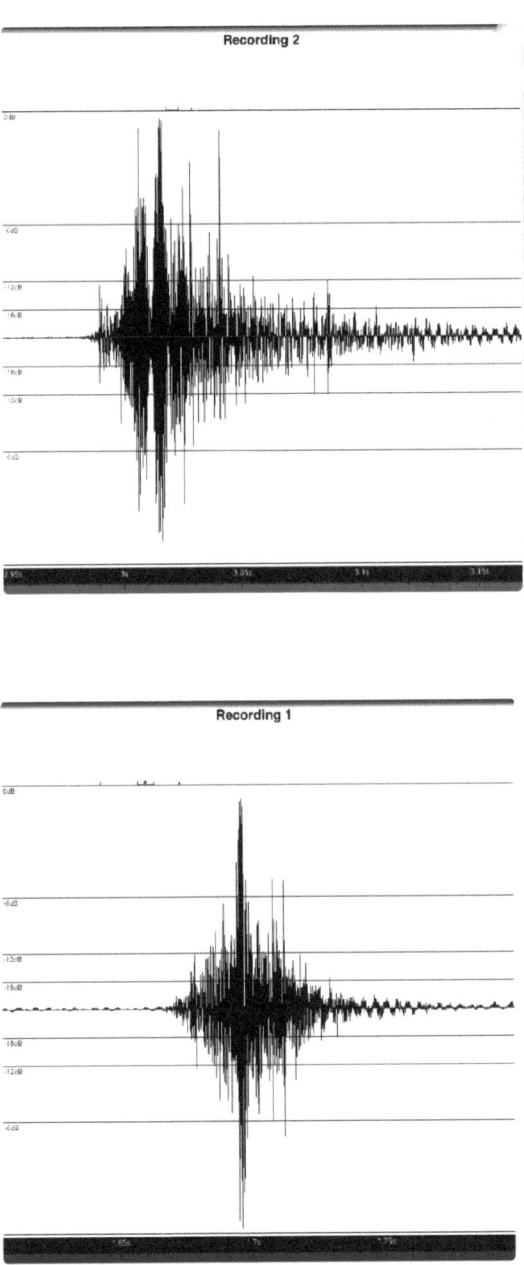

Recording 2

Recording 1

Above waveform is without and below with a fur strip attached to the string. Sound before the peak waveform is arrow related. All string vibration after arrow release is to the right of the peak spikes.

Both the fur strips and the application of rubber insert resulted in the reduction of vibration artifact. A clear dampening effect was seen with both applications. Arrows shot from a thicker string affected peak sound but resulted in no difference from silencer usage. Waveform artifact was reduced from the use of the applications, showing vibration dampening. Peak decibel levels were not affected but duration of the peak sound was reduced. It is concluded that the decibel levels from the string were not reduced but the overall duration of sound was. With the reduction of noise caused by vibration dampening, perceived pitch is also reduced.

During a primitive hunt, these changes would not affect the outcome of the hunt but it would affect the perceived feel and quietness of the bow. This study also showed that the effectiveness of these devises can be very important to a modern hunt. As shooting distances increase the amount of noise going towards the prey could affect the reaction of the animal. This is possibly even more dramatic if the string's lower pitch blends with the natural ambient noises of the hunting environment.

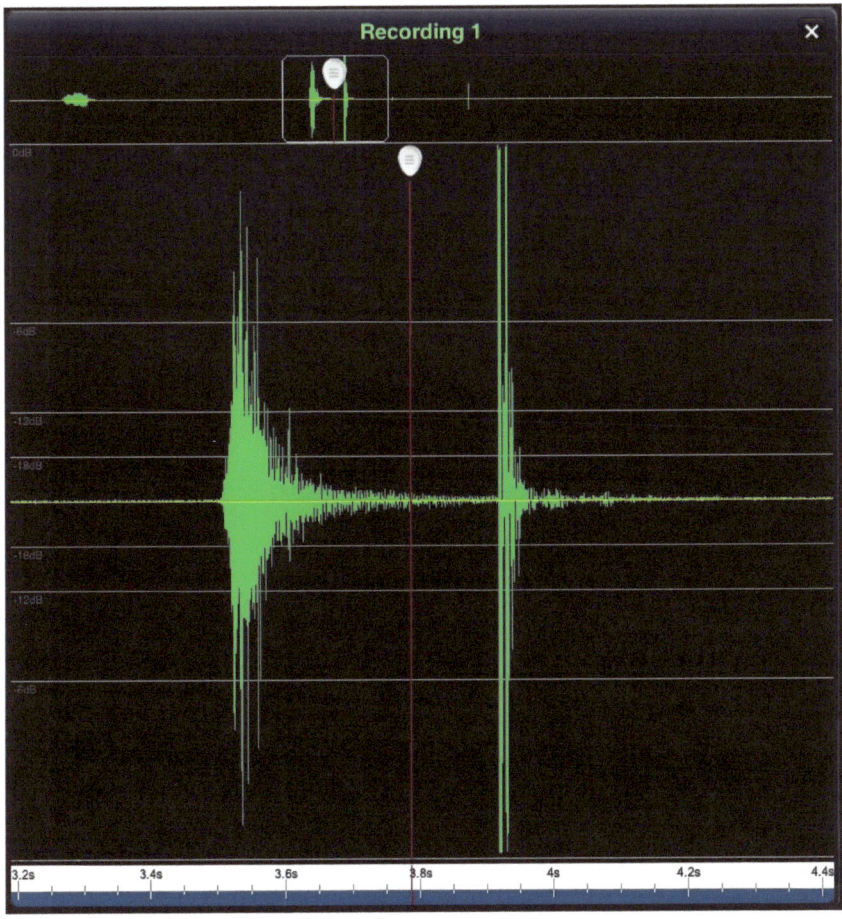

This is a full recording showing the sound created by the shot as well as the arrow impact on the target.

Takeaway in everyday language

Pictures show volume or loudness going up and down. Time in seconds is along the bottom. When comparing, you can see how the silencers took out the vibration artifact but did not reduce peak loudness.

This study is meant to address a common question among bow hunters. It does not address modern bows, arrows, or shooting distances. So for the primitive hunter when asked the question about the use of string silencers, I give you the following:

1. You still yelled but for a shorter amount of time.

2. The sound changed from a man's voice to a girl's voice so to speak (softened).

3. Possible softening of sound might make it blend better with natural area sounds.

4. It only should effect longer shots as you still yelled some.

5. Both fur strips and modern rubber silencers worked.

6. People like and buy bows that produce a pleasing sound.

Some people are not into primitive hunting with a bow and arrow. They may want to go even more primitive.

Dead rabbit! Using a Rock, Stick, or Knife?

When I was young and no larger than a jackrabbit, my family headed to a farm near Dickie, North Dakota to visit one of my relatives. As I remember it, fried chicken was on the menu so my aunt walked outside to the livestock area to obtain a couple of chickens. She picked up two large rocks and returned to the house with two chickens. My relative was rather old at the time and I don't recall that she actually wanted to chase chickens, thus the use of two rocks.

Hunting with a rock or stick in order to obtain a meal has been going on a long time. The question for this study arises as to what is the most effective method for obtaining a Midwest rabbit meal.

24" by 4" hand made by author out of Honey Locust.

A personal example of a hunting method for rabbits I have deployed is a stick in modified form. An opportunity occurred when I was deployed by the military to the Yuma, Arizona area. In my off-duty time, I was able to hunt the local antelope jackrabbits with a throwing stick. I used a branch fashioned from a local wood and halved it so that it had an aerodynamic advantage over its original form. This also increased its length to weight ratio. Time had to be spent observing rabbit behavior patterns, a learning curve containing hunting knowledge, and increasing throwing skills. It took around twenty misses before I got my first hit. The local cuisine was cooked with the skin left on and under the campfire. Of course I gutted it before cooking. A hot fast burning creosote bush was used to fuel the fire. This is not the traditional Midwest method for hunting or cooking rabbits that I grew up with but it worked.

The above story was to illustrate that different ecosystems require an adaptive, suitable, and productive hunting method. The story also shows a contrast with the real direction of this paper, which is the hunting of an eastern cottontail rabbit in a typical Midwest heavy brush environment. Research shows that a long, lightweight throwing stick is not an effective hunting tool when energy can be dissipated by environmental obstructions. Loss of striking power is taken away by vegetation like grass or brush before the stick can strike the prey. So what is the more effective weapon in this brush filled environment, a rock, stick, or thrown knife?

First, let's look at some stick and rock dynamics. Due to the brush factor around eastern cottontail rabbits, the heavier the brush, the shorter the stick. Looking through published literature, it is suggested a preferred size is 12 to 15 inches in length and a width of 2 to 3 inches of hardwood. This would give approximate weight of 1 to 1 1/2 pounds. A stick has an increased surface area to strike with as opposed to a rock, but as the stick gets larger it becomes harder to direct to the target. In both cases, speed of the throw is reduced, as the object gets heavier. A rock of 2 ½ to 3 inches in diameter weights approximately one pound and gives similar striking force numbers when compared to a stick. It was estimated at around 200 ft. pounds of energy were delivered at impact. Also the calculated variable of average speed can be within a range of 50 to 100 mph. This study picked an average HomoSapiens throwing speed of around 60 mph. A large adult can average 80 mph. but these numbers are from throwing a 5 1/2 oz. baseball. All factors ending up with the force generated being equal to mass times speed squared. Surface area differences between the two objects are as follows: A three-inch rock - calculated as circular cone having a surface area of 29.12 sq. in. While the stick – calculated as a 15 inch long circular cylinder with 155.50

sq. inches and a 14 inch length as 146.08 sq. inches respectively. This averaged as five times the striking surface area for the stick over the rock.

For a contrasting object to throw, I also wanted to look at the practicality of using a knife propelled at the small game prey. Knife throwing is a learned skill requiring practice to perform effective knife blade rotation, throwing a rock or stick is not in the same skill set. Hunting rabbits requires the ability to strike at unknown distances and with enough force to stun or kill. Missing the target increases the prey's learning curve making them more difficult to hunt next opportunity. In order to be realistic, one must be using the knife that would normally be carried when out in the field. Usually the one for carving wood and skinning of deer. It would be thrown at a rabbit when you first spot it hiding nervously in the brush. After much practice, this skill level was too high for the study comparison but data was left in for contrast. Distances would need to vary along with brush cover and uneven terrain. It was felt that a greater than 50% or better success rate was needed if you want to eat. Knife impact needs to stun or kill without the animal running off with your knife. Some people might be successfully, but time limited the learning of this skill to the level needed. Reasons for not throwing a knife may include that most knives are small and are not designed to be thrown, it is a valued object that has been passed down from father to son, and you get upset when the prey runs off with your knife.

In order to compare the differences between the variables of a rock, stick, or knife I enlisted two individuals to attempt to strike a 1.5L Pepsi soda bottle with sufficient force believed to be great enough to incapacitate or kill small game. The rabbit sized bottle was placed on its side, half filled with water to stabilize and simulate body weight. Ten throws were taken and distance related to hit percentage was recorded. The rocks were from 2 1\2 inches to 3 inches in diameter, sticks were unmodified, straight, and between 14 to 15 inches, 2 1\2 to 3 inches in diameter. The knife choice was a nine-inch SOG throwing knife. Throwing the rock and stick were childhood obtained skills while the throwing knife was not. With periodic practice over three months a skill level was obtained enough to add knife throwing to the comparison.

I would like to add an additional note, one of the testers was my older brother and during the first round of knife throws he used a 1.3L plastic Old Crow brand bourbon bottle and lost one of his knives after it bounced off the bottle into heavy grass. Drinking of alcohol was not involved.

So now you want to go hunting out in the woods.

Immune System and Medicine in Primitive Times

This topic can be difficult to discuss due to the emotional dynamic of wanting some type of spiritualism to supersede over logic. We as a general social group like to attribute status and great wisdom to the position of primitive medicine man or past native healer. This discussion looks at the use of medicine to cure diseases in past times. The purpose of this article is to question our point of view when addressing a medicine/primitive healer relationship. I believe that we as a collective attribute our current wealth of knowledge to physical healing principles in primitive times. This article is meant to double check our reasoning when asking the question, what really was the reality of medicine in the past?

I feel the majority of medicine in the past was a cause and effect relationship that required some sort of visual outcome. Otherwise, it would fall into a category of providing comfort to the patient as the disease process ran its normal course; modern treatment of the flu virus being an example of this. Past population isolation was a huge benefit to limit disease spread, and knowledge gained from cause and effect experimentation would limit exposure to harm like unsafe water.

Did you ever wonder, with all the diseases in the world, how mankind survived before hospitals and pharmacies? With over 650 common diseases able to inflict great punishment on humanity, it is a wonder we are still able to inhabit this planet. The top eight most common diseases alone can create 16,000 variations of attack leading the human host to destruction. Though we are not without a natural defense system, attacks to our system can come from bacterial, viral, fungal or parasite. All are dealt with by our immune system and defensive barriers such as our skin.

We don't catch a cold because we were attacked by a virus, but because we dropped our immune system defenses. Modern human do not get sick because of wet hair from a quick run in the rain. The illness is because we get chilled or stressed when standing in the rain and that has compromised our immune system. As time has passed, our immune system has not remained stagnant. Our genetic structure has been modifying it to accommodate the world around us. Natural selection kills off the weakest of the species leaving those most resistant to the environment able to reproduce. One of the several layers of defense that the immune system has can be to adapt to an attack. Barriers of protection are built up in anticipation of infection. These barriers can, over time be a thickening of the skin barrier, chemical level changes, or biological factors; all geared to minimize damage from a future attack.

One of the other weapons at our immune system's disposal are specialized cells, to use human warfare as a paradigm, these cells attack pathogens which need to elude the immune response to survive. A Special Forces like small group of hunter-killer T-cells search out pathogens to kill; also, in response to the attack, a biochemical flow of proteins specifically designed to assist antibodies in their action of killing pathogens – or bad guys. White blood cells are an independent unit or army of the basic immune response to the pathogen assault. Standard defense measures then adapt and overcome the enemy pathogen.

Understand that the genetics structure of the human organism has adapted to the existing environment in order to survive. The environment has changed and so has our immune system. We are structured for our current situation, type and amount of food, workload, sleep habits, and exposure to pathogens.

In the past, there was probably a higher death rate from disease, a lower life expectation, and a more robust immune system dealing with those elements of attack from that specific time period. With our modern lifestyle protecting us with advanced medicine, processed food, limited exercise and mechanical barriers such as air conditioned housing, we have lower death rates, higher expectations in life, and maybe have become more vulnerable to pathogens and are less likely to fend off a pathogen attack.

In the recent past, the most successful type of male provider was a John Wayne style stereotype male, able to kill, carry great burdens and build a log home. Current successful providers are Bill Gates stereotyped individuals of intellect and financial means. In the general survival of mankind sense, we have adapted to a changing world, physically as well as mentally and our immune systems have adapted to changes in modern stress levels.

Using diabetes as one disease process in epidemic proportion today, we can see the transition caused by quality of medical care. People died off from this disease in the past, never reaching reproductive age. Diets have changed dramatically, in sugar and salt content and in available variety. Availability of food is no longer the challenge it once was. As a longer life protected by medicine allows the diabetic gene to be passed on, obesity and lack of exercise have compounded the effects of the disease process having no similarity to past conditions.

The immune system was primitive mankind's primary defense against bacterial, viral and parasitic invaders, not food and medicine. Medicine men, or past healers, were taught by word of mouth, not by taking online classes. Past healers had few options compared to today's pharmacies. When it comes to a virus, one still treats the symptoms and does nothing to provide treatment against the virus. Fever is the immune system's way of killing the virus. Raise the core temperature beyond what a virus can live within, that is how the virus is killed.

Today's world is geared toward preventing exposure within a socially connected world of planes, trains and automobiles. Past civilizations had few social interactions outside a contained unit or small community. They had fresh air, exercise, and a cleaner diet. This would mean less food processing, no additives, and no cancer causing food coloring. In contrast, our modern time gives us dirt air, little exercise unless we join a fitness club, and a convenient processed diet.

We live twice as long in theory, and have access to vast amounts of medical education. Yet, many of us would love to drop the stresses of modern life and go back to a simpler time. If only our immune systems could handle it.

When looking at medicinal use of herbs and plants, are we attempting to treat modern conditions with modern knowledge for our current genetic structure?

The body will adapt to the stress we inflict on it given time. We can abrade our skin slowly over time and we develop calluses, but if we receive too much pressure too fast, we will blister. Primates have a lower survival death rate and longer life in captivity. In some cases, they live twice as long in the world of controlled environmental modern medicine. Humans as well as animals have cravings for specific types of foods and plants based on their internal needs. This is an inherited trait driven by urges without specific thought. As an example, we will crave certain foods containing elements such as salt that we may be deficient in. During illness, animals will eat bitter plants for stomach problems. I highly doubt that a complex pharmaceutical evaluation is going on here. It may be more just following the natural body urges. This is a natural occurring protective mechanism built into our primate level being for continued survival.

With our increase in scientific knowledge, store shelves full of food and medicine, and entire businesses based on natural plants that will prevent or cure every disease known to mankind, money may be a strong motivation for growth of this area. By combining marketing for money with 5000 years of Chinese medical history, we barely tap the knowledge base attributed to past medicine healers. Economic promotion of the discovery of some new use of an herb may not always be based in common knowledge from the past healers lexicon.

In my opinion, our genetic structure, immune system, and natural urges provided the core for our survival as a species. The vast majority of help from healers come from a few general-purpose plants common to that local environment. I feel patterns we see with small group primates holds true to the human species.

Rates of illness were related to change for the group. When illness strikes, the role of the healer was primarily providing comfort through a few beneficial herbs and a strong dose of spiritual support. Two pieces of information are needed for complete evaluation on this subject. They would be regarding isolated Amazon tribes with little to no contact to the progressive world. I would like to know age at death from what is considered normal old age and the number of occurrences in the physical use of medicine. Also what is the treatment of injury related to outcome? On the surface it appears all we really need is a few herbs to stop things like infection and some charcoal for a bad stomach, other than that let the immune system and spirit world deal with the rest. How often are we really medically treating a disease process to cure it resulting in a longer life?

In the past, times could have been simpler, it would seem so. But what about the spiritual aspect of the woods you hear so much about. How about if I include some food for thought on this subject as a last project.

You may be on video tape

Sometimes it is to your advantage to know a little background about the author before reading his material. It lets you know the piece of rock he is standing on or the point of view he is writing from.

I have a PhD degree in Healthcare Administration, went to college two more years after that in the medical field, and spent 17 years in the military. I grew up in the Midwest but over the years have lived on the West and East coast, in the South, and in several foreign countries. At one point in life I was so poor; I was adding weeds to my Ramen Noodles so it took away the hunger pains. At another point, I had so much money I was giving it away. I have lived with Kings and Peasants. Now that I am starting to feel old age creeping up on me, I am becoming very aware of just how much I don't know.

When you abuse an elder or steal something, do you feel you got away with it because no one saw you? I am here to tell you maybe you were on permanent video tape. Not only were people watching but the tape can be played back over and over again. Fortunate for you, the people watching are not judgmental, but instead love you, care about you, and worry about your choices. How do I know this? Using an old Native American story, the Thunderbird told me.

Who or what is this bird that is telling on you? Before I get to that, let me tell you of a story from my past. My mother asked me where I had been; I lied to her trying to stay out of trouble. She told me where I had been and what I had been doing. My mouth dropped as I stated, how did you know? She said a little bird told her. The next day my Grandfather saw me chasing the birds in the yard and asked me what I was doing. I told him I was going to catch the bird that told on me and kill it. He told me that the only way to catch the birds is by putting salt on their tails. I probably spent an hour chasing them around the yard with that salt shaker to the great pleasure of my Grandfather. I remember thinking I needed a five pound bag of salt to throw on them, which would be the only way I was going to get them. It was the start of my wonderful education.

Let me describe the Thunderbird as seen in the physical world.

This bird is larger than any you have seen before. It would have a huge, curved beak and eyes that glow like fire. Some say it is a pterodactyl left over from days gone by. Others say it has feathers like a pre-historic avian. Some of the stories tell of this great bird grabbing people and carrying them off. I think Hollywood has made several movies depicting this sort of image.

In the spiritual world it is described as a great bird that controls the rain, storms, hail, and lightning. It is large enough to carry a whale in is huge claws, powerful enough to cause floods, and with lightning coming from its eyes. It controls the good and bad effects of a thunderstorm.

The Thunderbird if it is real falls into the same category as Jesus or God being real. Jesus Christ in Christianity is seen as a physical man who had and still has powers far beyond that of mortal man. So how would we prove that Jesus is real? How about a combination of all the information around us, from the smallest blade of grass, the uniqueness of a forest, to the power of the thunderstorm. Is there a spiritual world is really what I am really asking?

For me, there are two worlds, a physical world and a spiritual world. As I see it, it is not in conflict with Christian beliefs or with traditional Native American beliefs. When you die in the physical world you pass into the spiritual world instead of no longer existing at all. There are a hundred variations and rules dealing with this by the religions of the world but the basic statement carries through as seen in near-death experiences.

Is the Thunderbird an allegory, symbolic of a greater truth, or maybe a spirit from the spiritual world who can affect the physical world? The Thunderbird controls lightning which is electricity, you know, the positive – negative electron transfer thing learned in school. Maybe it is too bad we as humans can't prove that the theory of electron transfer is real, yet we take it as a truth.

What if there is a spirit world and what if the spirit world could affect something earthly like the weather, would it then make sense to feel that when you steal something out of view from physical humans, would not the spirit people see it clearly and be very disappointed? This is what I learned from my parents, a little bird told them, and also some salt helps.

James Willer, PhD is a security expert who brings forth a unique blend of military, law enforcement and private industry protection experience with a deep understanding of the challenges faced by the Healthcare field.

Dr. Willer served as military police (MP) in the Army and Air Force branches of the United States armed forces for over 17 years. He has previous law enforcement experience and is certified as a police officer in the state of Nebraska. In addition to Dr, Willer's law enforcement and security background, he is also the founder of American Martial Arts. He is a 5th degree black belt in Taekwondo with a martial arts teaching certification from American Taekwondo Association in 1977.

www.ingramcontent.com/pod-product-compliance
Lightning Source LLC
Chambersburg PA
CBHW040813200526
45159CB00022B/559